ZERO KELVIN

Zero Kelvin

poems

Richard Norman

BIBLIOASIS
WINDSOR, ONTARIO

FIRST EDITION

Library and Archives Canada Cataloguing in Publication

Norman, Richard, 1981-
 Zero Kelvin / Richard Norman.

Issued also in electronic format.
ISBN 978-1-927428-45-0

 I. Title.

PS8627.O7633Z47 2013 C811'.6 C2013-902005-5

Edited by Eric Ormsby
Copy-edited by Tara Murphy
Typeset by Chris Andrechek
Cover Designed by Kate Hargreaves

Canada Council Conseil des Arts
for the Arts du Canada

ONTARIO ARTS COUNCIL
CONSEIL DES ARTS DE L'ONTARIO

Canadian Patrimoine
Heritage canadien

Biblioasis acknowledges the ongoing financial support of the Government of Canada through the Canada Council for the Arts, Canadian Heritage, the Canada Book Fund; and the Government of Ontario through the Ontario Arts Council.

PRINTED AND BOUND IN CANADA

Table of Contents

For Heidi

Desert

The tallest building ever built
begins to fill the desert up.

In the highest pane of glass
a ghost grows from a filament.

How beautiful in the evenings
when the scaffolding refracts
a hollow, shot-down sun.

See Venus shudder in the sky,
fruit of a jasmine tree.

From the penthouse suite the king surveys
the sand of sands
and then reloads.

It's in Ecclesiastes
from which we crib a verse each night.

From its future highest floor
men look back in time
and see the starlight scattershot
from far behind
full tumults of expanding dust.

Before he was told by a young Israeli
of the desert's beauty,
he believed in nothing.

He walked aimlessly in the desert city.
He didn't read the ancient verses.
A thin film coated the whites of his eyes.

Now the star is dimming on the silicate
and the horizon is Rothko's last dark work.

The arterial blood and dissipate brown,
crushed blueberry, and femur bruise.

He read from the book;
he became worshipful even.
The lines inside began to live.

Come with me to Lebanon, my spouse—
join me in the Levant.
Join me at the norias of Hama.

The ungodly groan of the wood
on the banks of the Orontes.

I once sat, looking out on the water before me.
I said, I will go up to the jasmine tree,
I will take hold of the boughs thereof:
I will be awoken at 7 a.m. by the front desk.
A tour of Krak des Chevaliers has been arranged.

Terrible as an army with banners,
I saw the sun march across the rocks.
I saw the turret above the hill.
In the valley saw the ghosts baked
and cracked like clay.

If all is futile under the sun,
we will find a way above.

There is a measure of blackness above
and there appear to be some holes.

It is written:
If the hole is feeding on in-falling matter,
the sun still rises on the giant desert—
dune sand finer than talcum,
almost liquid in this light.

Strange Days

The phone call is coming from inside the body.

The horror is
he or she is heard through the receiver,
but also somewhere even closer.

This stereophonic fear is new.

We walked on the surface of the frozen lake.
We stared at the chamber of the sky.
This was Siberia, far out in space.

Two sets of stars shone down
and scanned the lines of code we made.

There are people, you know nothing of them,
they form a plaque inside your veins,
show wonder to be the diminutive of fear.

Filament

At dawn we climb out of our canvas tents.
We look around and take deep breaths.
After the longest night of known grief
the sun knocks an electron out
of death, and shows the thing behind.

The spectral lines emerge and wrap
around the eyes from front to back.
The thinnest filament of presence
is there among the olive trees.
It starts to fill our field of vision.

Five hundred years have passed.
The horror vacui inspires grace.
Its reconstituted radiation
lights the worlds of our dreams.

Scan

The stars triangulate the cellphone call:
A woman captured in mid-song.

She bathes in an Italian bath
almost two thousand years ago.

The waves CAT-scan the cistern wall
and see the bones behind the brick.

The light reads each black bar of code
and makes a time and place you live.

The Kingdom of Kush

The sun evaporates the sky.
Only the desert remains
below the bright white blank.

It is strange to think that there are worlds in here,
above the core, that other things go on,
that they meet in this desert,
in this time stamped with evaporation.

Out of the dust in front of them appear
two black men on mules.
Inscrutable, unspeaking, they are approaching.
Perhaps 3,000 years will pass.

When we visited the Kingdom of Kush
in rags that hung from winnowed bodies
there was no welcoming committee.
Only a few equally bedraggled dervishes
who vanished on the spot.
We drank tea squatting on the roadside,
its flat flavour opening our pores and ducts.

The black people stood in the dust.
They were hidden in time in the dust.
We stood with camels at the well
dressed in rags, grit in the eyes.

There was the sorcerer dressed in terrible robes.
There was the sand stretching away forever.
Dust coated our faces. Quicklime covered theirs.

Truly, God is not hidden in the darkened sky
and the carcass of a camel soon means little
(said by a man claiming also to be the Mahdi).

The day grew long; nothing was done.
A stump from which grew lilies
stood in the nearby wadi.
The odd whiteness of what must be a hidden spring.
We put the fire out and slept.

We travelled south.
We went up in the mountains
to see another aspect of the stars.
We saw the green of the valley reflected in the stars.
One lit our faces,
vitreous humour of the heavens.
It was too beautiful to bear and we wept dew.

We slept closely in the verdant hills.
Truly, the living see further
in the light of this green star.

They simply laid together, minds blank,
as sand blew against their faces.
This was peace. To lie in the desert,
the long and pastel evening unrolling,
the world widening as we laid still.

From the desert to the river
there was green like the cloud around a star.

They were vine dressers from the mountains and from Carmel.
There were women somewhere in the dust-filled city.
They stood by mud-cake blocks, the water bore.
In my life I knew almost no one, but could grin and wave at strangers.

The strangers came from many lands,
all with a terrible purpose.
They slept, when they slept,
beneath the ash and cinder stars.

Here where the two Niles meet
the waters rushed away from us.
Unalmanacked winds bred and blew across the land,
and there was also the odour of dead cattle
and feces and spoilt food—
the odour of war.

I remember one evening in the desert:
On all sides were the mountains.
Everything sunk away but the sand beneath my tired feet
and that thing beneath it all
began to crawl
towards me from the blocked horizon.
All the branches of life that do not emit
appeared in the candelabras of lightning that gleamed down from the sky.

I remember leaving the tent
when the windstorm came
some time before dawn
the mind an empty room
doors and windows secured
and rain fell hot on the skin

of my shoulders and my back.
The sky was black and red.
More gore fell from the sky.
Everywhere around the sand was burning
but water came from the sky.

Lake Nyasa

A razor blade of starlight
makes a small tear at the hairline.

The sand-bound shore fills in
with the pink contagion of the dawn.

The scene is quiet, unbegun.
A glint reveals a jackknife, so you pull out a gun,

but then the beach is empty, and so you stand,
become the only thing within the field.

This pleasure means the sun is rising
and you're the only one it warms.

Dawn and the wide breeze,
the massive pastel vista of big blue Lake Nyasa,

and a young man swims into
the morning pool of induced blue.

Below him are the lives that are now full of water.
He treads the empty coma.

He smiles in the starlight as the water curls around him.
He knows he made himself a torus.

And other living things appear.
Even this soon after daybreak,

a conglomerate of flies,
a coiling funnel

which pours down itself,
expands, moves off of the horizon.

Every living thing
remote and programmed to come closer.

He treads blank water,
legs imitating an egg beater,

smoothing out the water with his hands.
The blue wide ripple, razure of oblivion,

spread out around him waving in the vacuum.
The whole world changes,

and the day declines,
sky draining into sun,

sun slowly churning blue into
a red that falls across the eyes.

City

Fall enjoins us, winter separates us,
but there is peace here in the city,
this entrepôt of amputees.
The gauze-like gaslights guide the weak.

Cold and lovely, there are lights
that trace the places people drink
and laugh and stare into the eyes
of other places.

Women grip the hardened hands
of distant men.
Men stare blankly at their feet.

The city floats. Below it is the sea
where no light shines.

Dig deep enough and water pours into
the hole from high above
the streets you walk along at dawn.

There is water above and water below.
Above the above water is where we cannot go.

I dreamt about the turmoil of the sea—
the figures falling down forever
or drawn up unexpected from the deep.
They wore disguises as in Bosch.

Do you remember?
The sky was the underside of the umbrella
we walked beneath.

The streets had emptied and the storm grew tall.

The war lit up the distant shore.

Vukovar

Near the shore a boat has bobbed for months.
Half-sunk, a broken jaw, its gunnels drink.
One night, some years ago,
a man walked out below the tower
and stared up at the black.
Without warning, the stars then opened fire.

It's the fact of day again.
Bombed houses gather birds
like yesterday's forgotten news.
The postman keeps on coming
though you long ago left town.

The doctor practices in Vukovar
when not in parliament in Zagreb.
He keeps a garden on the weekends—
the most beautiful for miles.
I sometimes see him standing in it.

A sunspot in the autumn afternoon
reminds him briefly of his son,
but the slowed beat of his heart
can't stand the effort.

The town left them behind.
Beneath the black and broken trees,
the soil clenches long-lost mines.
The sappers in their lawn chairs
on the Danube's banks
nurse rounds of noontime beers.

The town too has a beating heart.
It nearly brought each house here down,
pounding like a piston from the sky,
but these days are more calm.
The night comes with its hands held up,
not planning to ambush.
And the dawn's long diastole
floods the river's banks with light.

Pool

Visible matter made more visible
by staring through the moving water—

life and unlife mixed together,
threaded through, inseparable as light

moving through the medium.
Only the prism unweaves the colour.

The lines left on a sheet of glass
like traces of the people

waving there in place,
separable, peering down into the pool

from which you insist on staring up.

Goma

The odd call of a bird before the dawn.

Its two notes twist up in a helix
like vines competing in the humid green.

A tangled bridal mozzie mesh
hangs overhead.

The light switch doesn't work
and the shower is a bucket by the door.

The city, if it is a city,
sleeps in the depression,
a fog of gauze settling on leaves
which will not touch the ground.

The dark and complex forest shudders.
It vibrates in the mist.

The mountain overhead
is a poultice full of ash.

Old chimneys stand above the rooms.
Figures share a mattress laced with straw.
Their dreams are beams of light that hold the dust.

The single-circuit brains of goats and chickens
are posies on a ring of lead.
Their dreams are dark, emit no light.

At dusk there was no mist.
A mountain in a forest which is like the bottom of a pit.

Walking in a trance, among the women in the market,
the boys with idle, sidling eyes
slip into a mass of moving people.

The economics of contagion:
they catch you or are caught off guard.

One is grabbed.
The crowd surrounds him like a wound.
Viscous blood dangles briefly on a stick.

A red light lies beneath a thick, blank fog.
Everything is lost inside the dark place that exists.

At dawn the mountains overhead
swipe at the white relief planes flying in.
Each of them is lost.

How does one relieve
a city under siege
by its own ramparts?
asks a young female reporter,
in a city strewn with frozen magma,
among houses built of air-filled rubble.

One day the mountain
went down into the street

ignited the hospital and lean-tos,

ate the ancient motorcycles,
the carts and chikudus,

began to boil Lake Kivu from below—
bubbles of poison filling the sky ...

A similar substance as inside a star.

Light dangling on the particles within the cloud.

Light from another world
beneath the ashen strata.

Valve

As the sea anemone
that swallows water and propels away,
so the sea, the open valve,
the sea propelled away...

Impossible to say,
within the lexicon, a thing
about the tender floating out
into the whiteness,

the vast relief as night soon comes
and the stars, oil-lit lamps,
each a point of estimation
pushing us away...

Spring

The scent of spring like helium,
lighter than the air.

May in the new city
unfolds inside the soil,

puts out the buildings
and the people and the parks

it needs to sweep the rays of sun into its leaf.
The full green streets,
crab apple blossoms,

and this long lash-like breeze
that sweeps the city clean.

The young move through the blur.
It might be the belt that girds the air we breathe,

but no one knows for sure.
Likewise the couple posing in the park

beneath a massive furnace
in a photograph that lasts forever.

This specimen of light,
incapable of shadow,

which finds a way to trace
the duration of each figure.

Spring leaves the cities and the parks behind.
In excess, this light goes back.

Perdurable light
shifting blue to black,

meeting itself again
in the sudden
fillip of the star

that gilds a metal flower
opening in the fullness of space.

Harbour

This was Yokohama in the year 2006.
The trees still had their leaves
and night fell with an understated glamour
on the lane that lead up to the backlit harbour.

If air is turbulent, stars will scintillate—
if air is placid, they stay steady.
Particles fill the sky with matter
and the sea is also full of matter.

Everyone, please, all of you, enter.
The air here is all oxygen.
Veins of magma burn holes in the black,
mark out the temporary names.

The open-ended snowflake splits the eye
like lightning forced into a fishbowl.

Ocean

Drawn up gently through the sky,
darkness ascends from a winter dawn
like blinds collapsed into a single line.
The dark blue sky stares in.

Warmed breath injects into the air
a plume of sudden moisture
like iodine delivered into the blood
reveals a hidden structure.

Breath shows the crystal that becomes
the coursing ocean at Seabright,
reveals the waves of dark blue code
that hunt and feint beyond the break.

Winter makes us sleep in strange contortions.
Dreams shift with the body's heat
and each adds to the sum,
the land reseizing ocean

where you and I can live ...
We drift into the other's dream,
see waves pile up on one another,
whitecaps raised like flint-strikes.

The thing begins to take its shape.
The pines stand staggered by the light between them.
In a dream you often have
they topple one after the other,

a thousand rustic divers,
and leave each island bare.

The rocks roll down, add to the infill.
Each wave a smaller phrase

which the wind tries to decipher.
This is our little island
which gains a little ground
between each dusk and dawn.

On Winter Nights

On winter nights when the farmhouse doors are locked,
the moon is just a frozen ball of light
that keep its face locked to the Earth's.

She stays up late, reading by a fire
she takes great pleasure building every night.

People mistake wonder all the time, she thinks,
for fear. Outside, the river runs beneath the snow.
A slight scar indicates the water warm enough
to flow. She reads an atlas of the stars
and with binoculars will stand
before a blackened window
to look out of her patch of white
to everything that's ever been.

Her pupils dilate in the starlight.
She knows the pointlessness of sight.

She grins and tucks herself into her bed.
There are few pleasures like aloneness.

Cedar

The cedar is laid low by ice.
The night arrives to shear the sky.
This millennial winter proves
the pulse is now a depth charge.
Deep inside the sky it takes its silent sounding.

Meeting in mid-winter,
this friend pours a drink for me.
He toasts and pulls his wild beard.
He grins and starts to weep pure tears.
The drink I drink is black as ink.

In the volitional silence,
the edges of the drink
begin a slow retreat.
He is lying on his bed alone.
He is on a journey to the end of time.

Drinkers

Tonight the sky is scarred with stars,
the points of knives forever poised
above the bright, prone street of bars.

Drink dark water in the evening
and listen to the noise that pours out from the speakers.
So many men and women talk together
and their hearts apportion blood.

But he sits there unconvinced.
Given a drink, Tristan will speak to you of Iseult like a pimp.

They walk along the banks of snow,
the path lit up by littered salt;
a plow scrapes black the street again.

Compelled by what is possible,
the glass refilling at the tap,
a woman watches as they drink,
outlined eye held to a keyhole.
Then a key's inserted in the eye. One turn, it locks.

This was sometime a paradox,
but as they drink
the gravity accumulates,
the attention of strangers slowly develops.

The sharpness of feeling
coated by silicate.

The starlight floods the frozen sidewalks,
glares into the basement bars,
light moving from the core
to a bar in Granville Mall,
or off Queen Street.

There, replete with beer,
the drinkers sit,
features thickening—

more alive than the tableaux painted on the wall,
whose figures hold their drinks forever
unable to consume.
The night is a failure if it does not end.

Voices

In the station where we meet
the mobile phones ring all at once.
They take the calls, then clasp them shut.

The analog stranger, the digital lover,
are people plucked out of air,
long lists of light before they curved.

Seen through the window of a tiny restaurant,
a young man holds his chopsticks
like a fawn learning to walk.

A young woman smiles, continuing to talk.

We want the voice that comes out of the speaker
to be unbroken by the static,
but the substructure will not allow it.
So we get smaller to be certain.

To feel weak force upon your face
and see cities expanding at unnatural speeds ...

Time is too fast at this scale.
Just as the sculpture cut the fat off space
the voice becomes a single string.
It hangs before you and it waves.

Under The Volcano

Under the Volcano is a bar on 36th.
On these late Wednesday afternoons
colours saturate the street.
We order a round of Extreme Unctions
(double tequilas plus Tabasco)
and the fever peaks, the reins are blessed.
All the world's air conditioners
suck moisture from the streets.
People outdoors simply float away.
Their cars, apartments, souvenirs,
drift up and out, drawn in the gap.

This vacuum feeds us as we drink.
Outside, the skyscrapers create a lip
above which boil the clouds.
In an airplane high above the city,
I watch the screen in the headrest,
and see the lives lit up by X-ray lamps.
Below, the buildings are transistors
on an insulated board—
the streets are circuits and are gone.
The cloudburst as the plane banks right
reverses the last rites.

Qom

Freedom to walk down the day-drawn streets,
bask in the perihelion of the city,
count out each holy name.
Freedom to hear each song and what it stores.

The night bus arrives at the rest station.
Everyone exits, goes to the bathroom,
washes their forearms, their feet,
while over the speaker the recorded muezzin
recites the long part of the night.

A spectral line sung by a tenor,
a line of light between recesses.
And then the dawn arrives. We disembark
in the holy city of Qom
in the 25th year of Revolution.

The day is bright. Sunshine coats
the dried up riverbed, now a parking lot,
and the mullahs and their tasbih,
and every word that echoes in their heads.

Freedom to pull a Richard Burton
and slip into Fatima's shrine—
sit in the courtyard in late afternoon
reading the ancient code-book
trying to determine which symbols in it
once were wandering people.

The lines sink in the yellow page.
The ghost of dusk fills in the page.

Outside, the muezzin fills the dust
in the city's busy square.

It scintillates the air
and the names inside the air.

Seville

Spring in Seville, when the young mix rum and Fanta
and sit cross-legged by the fountain.
Reproduction here is the only philosophical question
and they squeeze into an alley of the Alcázar to ask it ...

In early evening, after the Corpus Christi floats,
the ancient moon fades into view.

More keenly than ever they feel
the blur that lingers over all their choices,
feel it sharpen in the alleys
as their answers flow into persuasion.

The young men and women
lean into one another
and the rum concocts
inside their blood
a blossom brighter than a girandole.

Killers

They first appear across the plain on horseback.

Dust rising from behind,
changing sunlight into resin
left clotting on the sand.

When they arrive in the small town
they kill their horses

and climb into two SUVs—

which burned out by the Rio Conchos
police find three days later.

At night, in the border cities,

in the humid early spring,
the street lights score the tar. The streets are wet.

The terrible aloneness of that sweat
cooling slowly on the skin
and the blades up there that circulate the air.
Just you and what the skin expels.
Ancient diver in a deep sea suit.

In the heat of the hundred houses,
which are prisons,
the squeezed-out lives,
the sudden evacuation not only of the bowels...

The gorgeous pool glows in the evening,
deep green elixir of chlorine.

On the bottom, dim leaves still mark
a pattern at midday—
shifting light that's never noticed.

The crowd suddenly goes quiet.
The band puts down their instruments.

The deeds of the killers,
lyrics of a long song sweated through on stage,

code to go to the washroom,
look in the third stall where you find the body,
blood inactive in the veins.

The self-selected samples all can speak;
that is the story's biggest bias.

The dark gulf like a field of asphalt
where nothing can be hidden.
The water endless and devoid of light—

out of which figures rise
like the hopeless photophilic plant
beneath a cellar door implores for light
which it cannot receive.

Or the distressed diver
who cannot tell which way is up.
Sun shining from the bottom of the sea.

Tsar Bomba

Bound in chains,
beyond distraught,
the star is brought
four klicks above
the Kola,
and executed
with a flaming bullet.

April

It is possible the real universe is smaller
than the one which scientists see.

Distant galaxies may be duplicated light—
circumnavigations regressively reflected.

Late one night in April, at a party with close friends,
I look across the table at my partner.
Dazzling, she holds a glass of wine.

It is possible what appear to be different galaxies

may be images of this galaxy
staggered through its lifespan
creating a limitless illusion.

Living Water

In the dream they always dream
a well of living water fills the fjord
and starlight spreads across the gorge.

The couple watch the small tree in their yard,
the trunk expanding,
roots seizing onto rock.

And then it fills their sleep.
Joint tenant dreams they never split.
Often dawn is grey, then yellow,
and they stand together in an orchard.

They stood together in a silent room—
this was not part of the dream—
and then night fell
much earlier than anyone expected.

The taut string of light
between two seas of black
snapped back. The elasticized end revoked its oath.
So many years have followed from that night ...

The long dream
replicating its stages across the night,
emitting neither star nor sign,
building a world without limit or consequence.

Deep, unchallenged dream
as the dead dream inside their graves.

The couple standing by the window,
granules dangling on indivisible threads
before the glass.

And light held on the leaves,
suspended in the recent rainfall,

the all-giving green
which also is not part of their dream.

Far From Here

In the evening by the phosphorescent pool,
long shadows waving in the moving water
and the company of paired-off friends
and the deep and energetic glow.
The star has looked upon me too.

I mean the one sunk deep into the ground.
The one which looks out from a well.
It's on another planet far from here,
a place where I will never go.

Lie back on the hotel bed.
A TV in a distant room
reports a major scientific breakthrough:
the Doppler effect observed in blood.
We love; we slide into a current of the past.

Solitaire

Paralyzed by ALS
a physicist sits at a seaside window,
sandwiches uneaten on a tray.
His guest has left the room to take a call.

A little song from long ago
streams out of the silkscreen of space.

Behind a pane of glass,
he looks out on the sea.
It may as well be methane.

Later, he plays cards.
He moves the cursor with a tube.
The wanderer crosses a field of baize.
The research questions are before him.

Will he get the aces out in time?
Or are they trapped and all the low cards too,
in the tall piles in the long row?

Event Horizon

They place in their hearts the event horizon.
A place beyond which happenings
cannot affect observers.
Light emitted from the inside
can never reach external eyes
and anything that passes through
is never seen again by you.

When observers think of death,
they imagine the nothingness
of dreamless sleep,
a revanchist darkness beyond recall.
Others, believing themselves
a type of information,
imagine they might be expelled in garbled form.

The Core

Once the drill bit has been built
big enough to dig it up
the tunnel provides us access
to the perfect place to hide—
the true source of the heat.

We tried to make a truce with it.
We stood on our two feet,
saw time-lapse of the trees pushed up,
felt strong and weak force on our face.

Colours began to fill the sky.
We watched light curl around the moon.

We built an interactive map
of every known star that's inside out
to find the flaw that made the formula berserk.

But it was a description
and would not change the thing about to happen.

Eye 1

How a nerve comes to be sensitive to light
hardly concerns us more
than how life itself first started ...
Any sensitive nerve can be rendered sensitive to light.
Any sensitive nerve can render light.
In the beginning the dumb word absorbed all light.
The silent world prevailed.
Now a faint string of pearls rings the ravishing word.
Now the eye makes spectres it sets upon
the half-lit bodies all around.

Eye 2

The way the eye turns away
temporarily
before the hush
as if to wish or usher in
disaster.
The witness
cannot help but stare.
The witnesses constrict.
It was commonplace once
to believe the eye emitted light
that it came to be night
when you chose to close them shut.

Experiment

It is a human urge—
to orbit backwards at great speed.
Experimentally, you do it
and then the crack of lightning,
the open-ended snowflake, splits the sky.
Just as the sculptor cut the fat off space,
you going backwards renders time.
Seconds drop like filings
when a magnet is turned off.

Operation

From high above the now night sky
a satellite begins to stare.

It has an eye that cuts right through.
More and more,

its circle is going elliptical
as it gets slightly older.

You stare into the cauldron of the sky,
induced to not be there,

and see the sky inside the body—
dark like the inside of a heart

and the lightning darker still.
Navy veins streaming down,

grounded in the spark
that makes the muscle start.

A machine washes out the blood
in the infinity of space.

Redaction

Every hole redacts a star.
A ghostly light surrounds each hole.
Most ghosts are just reflections back
from certain curves when they first bent.
Gazers trace their shape out of
the blackness that is all above.

Banquet

Like life on Earth, galaxies eat one another.
From close up a Caligulan banquet,
but dispassionately stately from a distance.
The Milky Way's neighbour, Andromeda,
currently devours one of her slaves.
When you attend, you watch with fascination
as she towels off her maw
and spits the remnants out
into the vastest vomitorium.

More than a dozen clusters
lay scattered around Andromeda,
cosmic remains of vast past banquets
and the preceding emetophilia.
Prophets who know the scientific method
believe our galaxy and its neighbour will eat each other
three billion years from now.
Social mobility being what it is,
slaves may then be emperors.

Solar Sail

Look out upon the surface of Titan,
the sea of methane impossibly unfrozen.

Weigh the weight of the fire,
or the blast of the wind,
or bring back a day that is past.

Be the prophet who gazes through the speculum
and sees an image like a face.

What's in the polished stone
is the same blackness that stores mnemonic static.

Look through a telescope darkly
at that old time, the face inside the static.

A high priest by the name of Eric Demaine,
youngest professor at MIT,
will adapt the map fold to a solar sail.

Watch it enter its first orbit,
see it slowly open,
our chrysanthemum in space.

Theology

Objects crossing or approaching the orbit of Neptune ... are given
mythological names associated with the underworld.
—"How Minor Planets Are Named," International Astronomy Union

An image appears in the crafted glass.
The same image that will shrink to fill a contact lens.
The same horror in an instant

of losing irretrievably an heirloom.
It's only natural stars recede
from the expectation of a billion gazes.

But everything is stored. The night returns restored
projected from the data.
Behind the screen the algorithm

(soon to graduate to etiquette)
reveals the folk inside the medium.
These women photograph themselves,

upload their dust into a cloud.
Seeded, these banks of clouds will fill—
each particulate of dust, each pearl congealing.

Theology, the study of dark matter,
conclusively has proven
the well of hell is zero Kelvin.

Movement ceases,
molecules foetally curl into themselves.
And at the lowest circle of our galaxy

a black hole squats.
O wondrous Goatse of another realm!
Radio source,

mass of four million suns,
beams out pure revelation.
Cults worship at its altar.

The faithful pray:
Do not leave your house—
sit quietly and listen.

An LED illuminates
the ether in the vitrine.
And models show the diodes rapidly receding

and the backlit screen expanding,
and the transudation,
and something dug up from deep within

that will not act and will not leave,
a thing that makes a truce with space,
a relic of the underworld.

Patmos

After Nikolai Morozov

The thorn trees in the terraced yard.
The little place, below the little sun.

The gleaming face, beneath the girandole
of bursting stars.

Look at the figures in the sky.
Look at the horsemen riding there.

All has been assigned on this last day of life.

Acknowledgements

The author is grateful to the editors of the following journals where some of these poems first appeared: *The Malahat Review, The Puritan, CV2, Existere, Saint Ann's Review,* and *QWERTY.*

About the Author

RICHARD NORMAN lives in Halifax. *Zero Kelvin* is his first collection.